FLORA OF TROPICAL EAST AFRICA

TECOPHILAEACEAE

Susan Carter

Perennial herbs with corms or tubers. Leaves petiolate, sessile or amplexi-caul, linear to ovate-orbicular, decurrent to cordate. Inflorescence racemose, pseudo-racemose or paniculate, or the flowers solitary. Bracts sometimes minute or absent. Flowers pedicellate, ☿, regular or slightly irregular. Perianth-segments 6, free or forming a short tube, entire, spreading or re-flexed. Stamens 6, inserted on the perianth-segments at the throat of the tube, all perfect or 2 or 3 replaced by linear staminodes; anthers dithecous, sometimes connivent, with the base sometimes produced into a spur or sac; dehiscence by an apical pore, usually slightly introrse. Ovary semi-inferior, rarely almost superior, trilocular; ovules 2–∞ in each cell; placentation axile; style filiform; stigma minutely trifid. Capsule trilobed, dehiscence loculicidal; seeds small and numerous, or a few large ones, rarely only one developing, smooth or warted, rarely with spongy cap; endosperm present.

A family of four genera from the New World including the type genus *Tecophilaea* Colla, and three from tropical and South Africa.

Tubers in a vertical series; leaves basal; inflorescence
    racemose or pseudo-racemose . . . . . 1. **Cyanastrum**
Tubers in a cluster; leaves cauline; flowers usually solitary,
    arising in the axils of the leaves . . . . 2. **Walleria**

1. CYANASTRUM

Oliv. in Hook., Ic. Pl. 20, t. 1965 (1891); R. E. Fries in Wiss. Erg. Schwed. Rhod.-Kongo-Exped. 1: 225 (1916); R. T. Clausen in Bailey, Gentes Herb. 4: 293 (1940); Carter in K.B. 16: 190 (1962)

Tubers depressed-globose, superposed to form a vertical series; roots thick and fibrous. Cataphylls 1–3, scarious, sheathing. Leaves basal, petiolate, rolled in bud; leaf-blade ovate and cuneate, to ovate-orbicular and cordate. Flowering stem sometimes appearing before the leaves, naked or with 1–2 distant scarious bracts, the lower usually sheathing. Inflorescence pseudo-racemose; bracts large and ovate to linear, or minute and sometimes absent; bracteoles sometimes present and then usually subtending a flower. Perianth-lobes in two series but joined at the base to form a very short tube. Stamens all perfect; filaments filiform; anthers linear or oblong, with an apical terminal or introrse pore. Ovary semi-inferior, deeply trilobed; ovules 2 in each loculus; style longer than the stamens. Mature seeds 1–3, forming a 1–3-lobed capsule; seeds rather large, oval, smooth, with a large spongy cap.

A genus of four species, with the type species, *C. cordifolium* Oliv., from West Africa (Nigeria to Gabon), and the other three from eastern tropical Africa. These latter have all been collected rather infrequently from parts of western and southern Tanganyika.

Habitat notes are scanty, but it seems these species are found particularly in shady rather moist places in the transitional zone between deciduous woodland or thicket and riverine forest or rain-forest, also perhaps along river-banks. *C. johnstonii* Bak. var. *cuneifolium* S. Carter however, was found in open deciduous woodland. Further observations on the apparently rather unusual ecology of these species would be interesting.

Leaf-base cuneate to cordate; young leaves rolled separately from the flowering stem; perianth usually blue; apical pore of anthers terminal:
Inflorescence ebracteate and rarely with 1 or 2 filiform bracteoles; nodes of inflorescence 1-flowered . 1. *C. johnstonii*
Inflorescence bracteate; nodes of inflorescence mostly 2-flowered . . . . 2. *C. goetzeanum*
Leaf-base decurrent; young leaves rolled round the flowering stem; perianth white; apical pore of anthers introrse . . . . 3. *C. hostifolium*

1. **C. johnstonii** *Bak.* in F.T.A. 7: 336 (1898). Type: Zambia, hills between Lakes Tanganyika and Nyasa, *H. H. Johnston* (K, holo. !)

Tubers up to 2·5 cm. across. Cataphylls 2–3, the outer about 3 cm. long, the inner 6–9 cm. long. Leaf-blade ovate, up to 15 cm. measured along the midrib, 12 cm. wide, cuneate, rounded or cordate, with lobes 3·5 cm. long; petiole up to 15 cm. long. Flowering stem up to 30 cm. high, appearing before the leaves. Stem-bracts 1–2, the lower sheathing, up to 5 cm. long, the upper usually free, linear, up to 4 cm. long. Nodes of inflorescence 5–20; secondary peduncles 1-flowered, the lower ones rarely 2-flowered, with pedicels 4–10 mm. long; floral bracts absent; bracetoles absent, or represented by minute scars, or when present (very rarely) minute and filiform. Perianth blue; lobes 12–18 × 3–6 mm.; tube 1·5 mm. long. Anthers yellow, linear, 5–7 mm. long; pore terminal; filaments 3–4 mm. long. Style 10–14 mm. long. Capsule-lobes oval, up to 10 × 7 mm.; seed whitish, rounded and compressed to 7 × 5 mm., with a brown, reticulated, spongy cap, 7 × 6 mm.

var. **johnstonii**; S. Carter in K.B. 16: 194 (1962)

Leaf-base distinctly cordate.

TANGANYIKA. Buha/Kigoma Districts: 30 km. N. of Kigoma, Gombe Stream Game Reserve, 4 Nov. 1960, *Goodall* 107! & Mkenke stream valley, 15 Mar. 1964, *Pirozynski* 562! & Kasakela, 17 Nov. 1962, *Verdcourt* 3344!
DISTR. T4; Congo Republic (Katanga) and NE. Zambia
HAB. Usually in transitional areas between deciduous woodland and riverine forest, in shady and often rocky places; 820–1500 m.

SYN. *C. verdickii* De Wild. in Ann. Mus. Congo, sér. 4, 1: 5 (1902). Type: Congo Republic, Katanga, Lukafu, *Verdick* 275 (BR, holo. !)
C. hockii De Wild. in F.R. 11: 517 (1913). Type: Congo Republic, Katanga, Elisabethville, *Hock* (BR, holo. !)

NOTE. The type-specimen of *C. johnstonii* appears to have been " lost " in the Kew Herbarium until now. Consequently authors have followed Baker in describing the species as having deciduous bracts, but I can find no evidence of their presence on the type-specimen. The number and size of flowers in an inflorescence is very variable, so I do not hesitate to follow Fries in reducing the ebracteate species *C. hockii* to synonymy and adding also *C. verdickii*.

var. **cuneifolium** *S. Carter* in K.B. 161: 194 (1962). Type: Tanganyika, Mpanda District, *Richards* 11638 (K, holo. !)

Leaf-base cuneate or rounded, not distinctly cordate.

TANGANYIKA. Mpanda District: Kapapa Camp, 20 Oct. 1959, *Richards* 11638!
DISTR. T4, known only from the type-gathering
HAB. In deciduous woodland on sandy gritty soil; 1050 m.

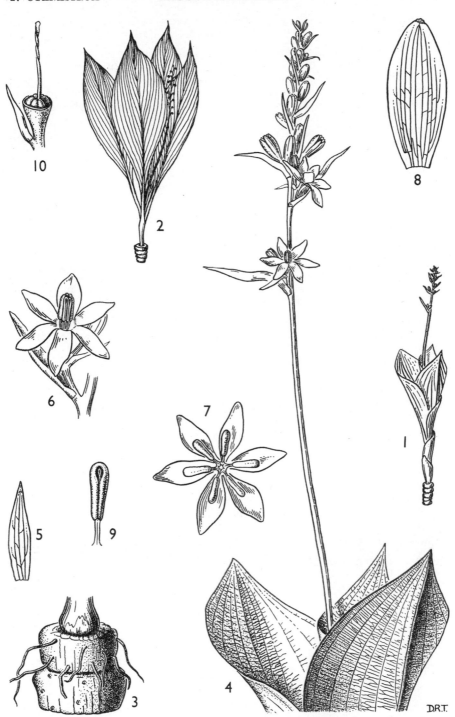

FIG. 1. *CYANASTRUM HOSTIFOLIUM*—**1,** plant in flower, × ⅙; **2,** plant in fruit, × ⅙; **3,** rootstock, × 1; **4,** inflorescence, × 1; **5,** bract, × 2; **6,** biflowered secondary peduncle, × 2; **7,** flower opened out, × 2; **8,** tepal, × 4; **9,** stamen, × 4; **10,** pistil, × 4. 1, 2, 4–10, from *Gomes e Sousa* 879; 3, from *Faulkner* P.330A.

2. **C. goetzeanum** *Engl.* in E.J. 28: 359 (1900). Type: Tanganyika, about 120 km. E. of Iringa, Lofia R., *Goetze* 438 (B, holo. !)

Tubers up to 2 cm. across. Cataphylls 1–2, the outer ± 2 cm. long, the inner 7 cm. long. Leaf-blade ovate-orbicular and deeply cordate, up to 8·5 cm. long measured along the midrib, 10 cm. wide, with lobes 5 cm. long; petiole up to 18 cm. long. Flowering stem up to 26 cm. high, appearing as the leaves begin to unroll. Stem-bracts 2–3, the lower sheathing, up to 4 cm. long, the uppermost usually free, up to 2·5 cm. long. Nodes of inflorescence 3–6; secondary peduncles nearly all 2-flowered but sometimes only one flower developing, with the pedicels 4–8 mm. long; floral bracts linear, 2–8 mm. long. Perianth blue (or white, *fide* Engler); lobes 10–12 × 3–4 mm. Anthers linear, 4–5 mm. long; pore terminal; filaments 1·5 mm. long. Style 6–8 mm. long. Capsule and seeds not seen.

TANGANYIKA. Ulanga District: Ifakara, Machipi, Nakatimbo, 11 Dec. 1959, *Haerdi* 391/0 !; Iringa District: about 120 km. E. of Iringa, Lofia R., Jan. 1899, *Goetze* 438 !
DISTR. T6, 7; not known elsewhere
HAB. Little known, but does occur near streams, probably shady margins of riverine forest; 380–600 m.

NOTE. This rare species is very similar to *C. johnstonii* but can easily be distinguished by its shorter and, in comparison, wider leaf (which is much more like that of *C. cordifolium* Oliv.), the presence of small persistent bracts and its smaller fewer flowers which are usually produced two to a node.

3. **C. hostifolium** *Engl.* in E.J. 28: 358 (1900). Types: Tanganyika, Uluguru Mts., *Stuhlmann* 1894 & between Ukutu [Khutu] and Uhehe, *Goetze* 394 (both B, syn.)

Tubers up to 2·5 cm. across. Cataphylls usually 2, the inner up to 7 cm. long, the outer up to 3·5 cm. long. Leaves rolled round the flowering stem, elongating when the plant is in fruit, with petioles up to 12 cm. long; leaf-blade ovate to ovate-lanceolate, 15–30 × 4·5–9 cm., mucronate, cuneate. Flowering stem up to 35 cm. high. Sterile bracts rarely present, lanceolate, up to 5 × 1·5 cm. Nodes of inflorescence up to 25; secondary peduncles 1- or the lower often 2-flowered, with the pedicel up to 10 mm. long; floral bracts linear to lanceolate, the lowest up to 30 × 6 mm., deciduous at the fruiting stage. Perianth white; lobes 10–14 × 3·5–5 mm.; tube 1 mm. long. Anthers yellow, oblong, 3–4 × 1 mm.; apical pore introrse; filaments about 2 mm. long, broadened at the base. Style 4·5–6 mm. long; ovary deeply 3-lobed. Capsule and seeds not seen. Fig. 1.

TANGANYIKA. Morogoro District: Turiani, 22 Nov. 1955, *Milne-Redhead & Taylor* 7356 !; Ulanga District: Ifakara, Funge, 9 Dec. 1959, *Haerdi* 223/0 !; Newala District: 50 km. E. of Newala, Ruvuma valley, 15 Jan. 1959, *Hay* 27 !
DISTR. T6, 8; Mozambique
HAB. Little known, but mostly recorded from transitional areas between deciduous woodland or thicket and riverine forest or lowland rain-forest, usually in shady places; 0–680 m.

SYN. *C. bussei* Engl. in E.J. 38: 58 (1905). Type: SE. Tanganyika, "Makondeland", Seliman-Mamba, *Busse* 2667 (B, holo. !)

NOTE. Plants of *C. hostifolium* grown at Kew left no doubt that *C. bussei* is synonymous, when their leaves elongated and expanded after flowering to show the petiolate leaf described as the distinguishing feature of *C. bussei*.

2. WALLERIA

Kirk in Trans. Linn. Soc. 24: 497 (1864); S. Carter in K.B. 16: 185 (1962)

Tubers clustered, globose, with fibrous roots, 10–15 cm. underground. Stem ridged, smooth, scabrid, or armed with recurved prickles. Leaves all

FIG. 2. *WALLERIA MACKENZII*—**1**, tuber, × 1; **2**, flowering stem, × 1; **3**, flower, × 2; **4**, perianth opened out to show stamens, × 2; **5**, tip of anther, × 10; **6**, fruit, × 1½; **7**, fruit, × 1; **8**, seed, × 4. 1–5, from *Milne-Redhead & Taylor* 8520; 6, 7, from *Bullock* 2200; 8, from *Richards* 5025.

cauline, alternate, sessile or amplexicaul, linear to ovate, the midrib on the under-surface sometimes scabrid or armed with prickles. Flowers axillary, solitary; pedicels long, smooth, scabrid or prickly, erect or recurved; bracteoles rarely subtending a second flower. Perianth-segments forming a very short tube; lobes spreading but after anthesis folding up to enclose the ovary. Stamens all perfect; anthers linear, sometimes connivent at the tips, with an apical terminal or introrse pore; filaments filiform. Ovary almost superior; ovules about 8 in each loculus; style a little longer than the stamens. Capsule globose; seeds 2–3 in each cell, ovoid, dark red-brown and shiny, densely warted.

A genus of three species from tropical and South Africa.

W. mackenzii *Kirk* in Trans. Linn. Soc. 24: 497, t. 52 (1864). Type: Malawi, Manganja Hill, *H. Waller* in *Kirk* (K, holo. !)

Tubers 2–4 cm. across. Stem smooth or sometimes scabrid, up to 80 cm. high. Leaves of variable shape from ovate to linear-lanceolate, with the upper leaves more lanceolate, from 7 × 0·5 cm. to 9 × 2 cm. and 6 × 2·5 cm., sessile. Pedicels up to 5 cm. long; bracteoles at about the middle of the pedicels, linear to lanceolate, rarely up to 2 cm. long, sometimes subtending a flower with a pedicel up to 2·5 cm. long. Perianth purplish-blue to grey-blue or mauve; lobes 14–19 × 4–6 mm. Anthers deep blue with yellow tips, free, 8–11 mm. long; apical pore terminal; filaments 1–2 mm. long. Style pale blue. Capsule up to 2 cm. in diameter; seeds 5 mm. long, 3 mm. across. Fig. 2.

TANGANYIKA. Tabora District: Ushetu, 10 Jan. 1955, *Akiley* 5030 !; Songea District: 1·5 km. S. of Gumbiro, 24 Jan. 1956, *Milne-Redhead & Taylor* 8520 !; Lindi District: near Ruponda, Namanga, 1 Jan. 1949, *Anderson* 281 !
DISTR. **T**4, 7, 8; Angola, Congo Republic (Katanga), Zambia and Malawi
HAB. Deciduous woodland and bushland on red sandy loam soils, sometimes edges of grassy depressions with impeded drainage and on ant-hills; 300–1500 m.

SYN. *W. angolensis* Bak. in Trans. Linn. Soc., ser. 2, 1: 262 (1878). Type: Angola, Huila, *Welwitsch* (BM, holo. !, K, iso. !)

NOTE. This species is very variable in leaf-shape, and the type-specimen of *W. angolensis* shows one extreme with its very narrow almost linear leaves. However, the many specimens seen show a complete range between this and an ovate shape, some large plants possessing both shapes, with the upper leaves linear-lanceolate.

INDEX TO TECOPHILAEACEAE

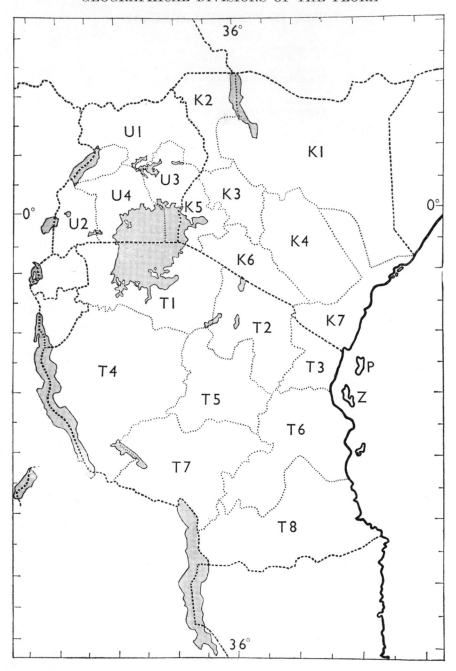